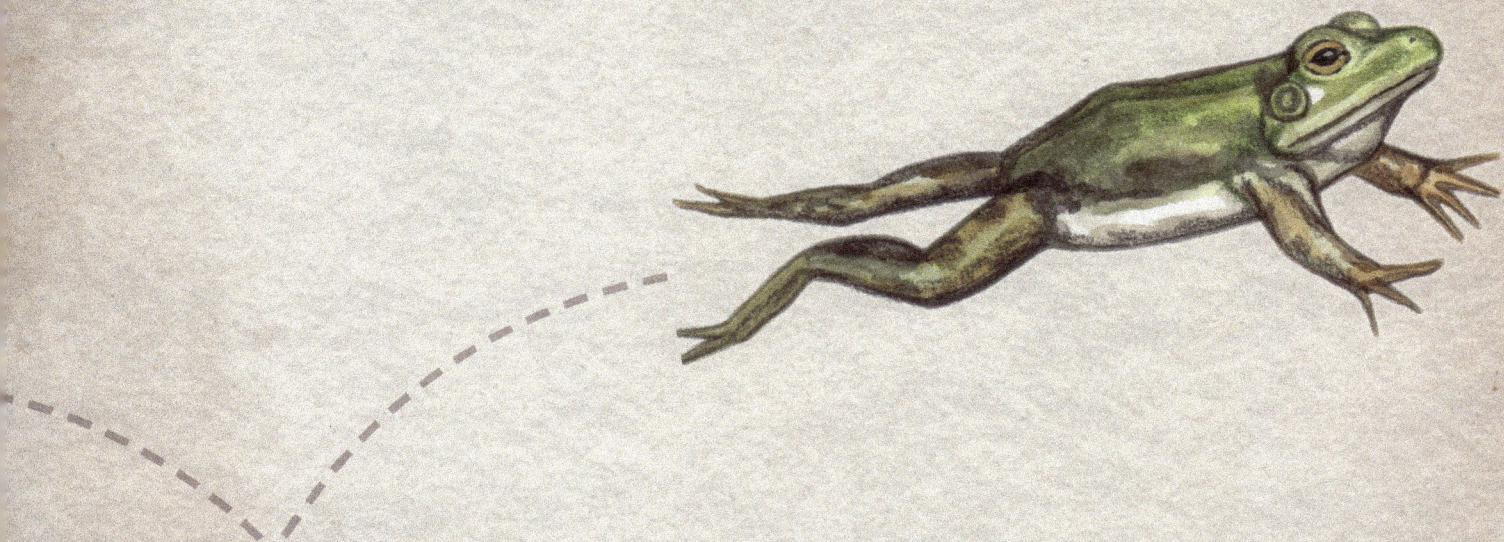

This book belongs to:

Preface:

This book began in 2021 as a 100 day self-run research project to learn about amphibians that interested me. I grew up surrounded by 15 acres of field land, wooded area, ponds, and streams which allowed me to immerse myself in nature daily. I found myself drawn to the small animals and I remember my hands-on experiences with amphibians so well. Those memories have now sparked my interest to dive deeper into the topic as an adult. In the best way I know how to learn, I proceeded to illustrate these amphibians, their environments, and the letters of the alphabet inspired by characteristics of the individual species.

For 100 consecutive days, I shared an illustration each day online as I searched the internet about each species. Those following along guessed what species was coming next based on the letter and environment illustration. With my main gig being illustration, I had no previous education about Amphibians.

To ensure I collected accurate information along this journey- many skilled peer reviewers volunteered their time, and contributed their knowledge of amphibians. I would like to dedicate this book to them.

Natalie M, Jenna V, Victoria R, Melissa F, Ryker H, Zach B, Amelia G, Anthony S, Austin R, Faith K.

Copyright Erica Bradshaw, 2022.

All rights reserved. No part of this book may be reproduced in any form without permission in writing from the author.

Bradshaw, Erica
Alphabet of Amphibians.
Kalamazoo, Mich.: To Draw Attention Publishing, 2022.
46 pages illustrations
1. Discovery 2. Animals 3. Adventure
BRAD
ISBN: 978-1-7327659-1-7
Book Design and Layout: Erica Bradshaw
Kalamazoo Mich.
Printed by LSI

Author and Illustrator Erica Bradshaw

Alphabet

of

Amphibians

an illustrated field guide

Written and Illustrated by Erica Bradshaw

A is for Amphibian!

What is an amphibian?

Amphibians are cold-blooded animals that require a moist environment to survive. Since amphibians breathe and hydrate through their skin, it is important they live in a habitat with water readily available.

Amphibians are also classified as having two main stages of life:

1. A gill breathing underwater larval stage, almost like a fish!

2. A terrestrial stage where they live on land breathing air through their lungs like you and me.

Some amphibians live a semi-aquatic lifestyle where they can live both on land and in water. You most often see this with frogs!

Types of amphibians:

Frogs, toads, newts, salamanders, and caecilians

Over 8,000 species of amphibians exist in the world. 90 percent of these are frogs. Only about 300 of them are found in the United States. Throughout this book we will learn about 26 of them from around the world!

A is also for Axolotl!

In the wild, an axolotl is a special, rare type of salamander that permanently lives in a larval stage.

Most salamanders will grow up and be able to transition between land and water. However, an axolotl's body doesn't quite reach the point where they can live on land, so they will live their entire life in water.

Due to this premature halt in metamorphosis, the Axolotl still has its dorsal fin on its tail, similar to a tadpole.

Ambystoma mexicanum

Living life underwater axolotls have external gills. They are the fluffy looking growths from the sides of their head.

Some axolotl's have light, and translucent skin like above; but black, and mottled brown varieties are most common in nature!

Axolotls are only found in one place on earth: The Xochimilco Lake Complex near Mexico City.

Critically endangered & nearly extinct: Their population is declining in the wild due to pollution, water drainage, and large fish being introduced to their home lakes.

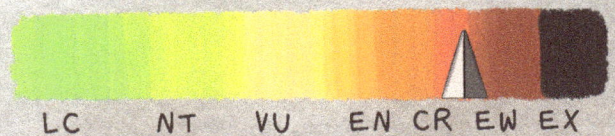

LC   NT   VU   EN CR EW EX

The Life Cycle of an Amphibian
comparing frogs and salamanders

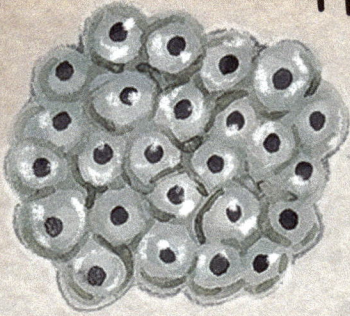

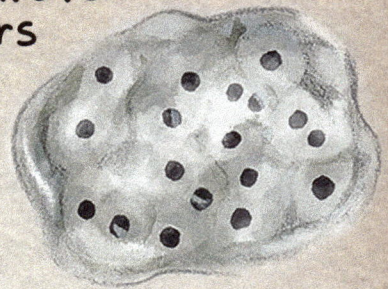

Clumps of eggs
are laid in water

Salamander eggs have
a jelly like protective
outer layer.

Tadpole

Tiny tadpoles or
larva emerge from
eggs

Larva

Tadpoles and Larva
begin developing limbs,
and gills.

Their back limbs
develop first.

Their front limbs
develop first.

Tadpoles and Larva develop
four fully formed limbs, but
retain tail fin.

Frog's will eventually lose
their tails to help them
move easier on land.
Salamanders retain
their tail.

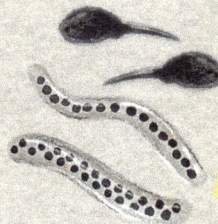

Whats the difference
between Frogs vs.
toads?

Life cycles are nearly
the same!

Toad eggs are formed
in lines, and TOAD
poles (ha!) tend to be
solid black in color.

Conservation Status of Amphibians
as explained by the International Union for Conservation of Nature:

The IUCN is a global organization responsible for the protection of nature by researching species and putting together practical plans to help conserve them. The IUCN conservation status records whether an animal or plant species is threatened by extinction. The conservation status is based on scientific information by specialist groups.

Throughout this book you will find a scale marking the conservation status of each amphibian. The scale and its meaning is outlined below:

LC: Least Concern. A species is widespread and abundant

NT: Near Threatened. A species likely to qualify for threatened category, perhaps very quickly depending on local development projects such as rain forest logging.

VU: Vulnerable. A species faces a high risk of extinction in the wild.

EN: Endangered. A species faces a very high risk of extinction in the wild.

CE: Critically Endangered. When a species faces an extremely high risk of extinction in the wild.

EW: Extinct in the Wild. When a species survives in captivity, cultivation, or has an extremely small population.

EX: Extinct. After exhaustive surveys, the last known individual has died.

Leading Contributing Factors

Habitat Loss
Urban development is filling in streams and ponds as well as clearing natural habitats for housing and roads.

Climate Change
Dramatic shifts of temperatures, wildfires, and drought alter conditions needed for amphibians.

Chytrid Fungus
A water-dependent fungus widely infects amphibians skinmaking them unable to breath.

B is for Blue Poison Dart Frog!

Poison dart frogs are tiny, but mighty! These brightly-colored tropical frogs are known for a natural poison in their skin. Their poison skin is used as a defense to paralyze, or even kill predators!

Some species of dart frogs' poison is so strong it has been used for centuries by Indigenous cultures to coat the tip of their blow darts before hunting which inspired the name "Dart Frog!"

Blue poison dart frogs live on the rain forest floor. They usually live under rocks and moss near streams, they can sometimes be found high up in trees.

The frogs poison is created by highly poisonous ants that the frogs eat in the wild! Once in captivity, they are no longer toxic.

Blue poison dart frogs live in isolated areas of the rain forest in Suriname, and northern Brazil.

Dendrobates tinctorius
"azureus"

Least Concern: Their population is stable.

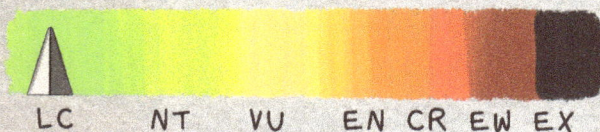

LC   NT   VU   EN CR EW EX

C is for Clown Tree Frog!

Clown tree frogs are known for their vibrant clown-like color combinations! Previously considered to be part of another species, these tree frogs are often misidentified.

In the wild, these tree frogs inhabit the tree top canopy of the Amazon rain forest. When it is time for breeding, they will make their way down to forest pools.

Dendropsophus leucophyllatus

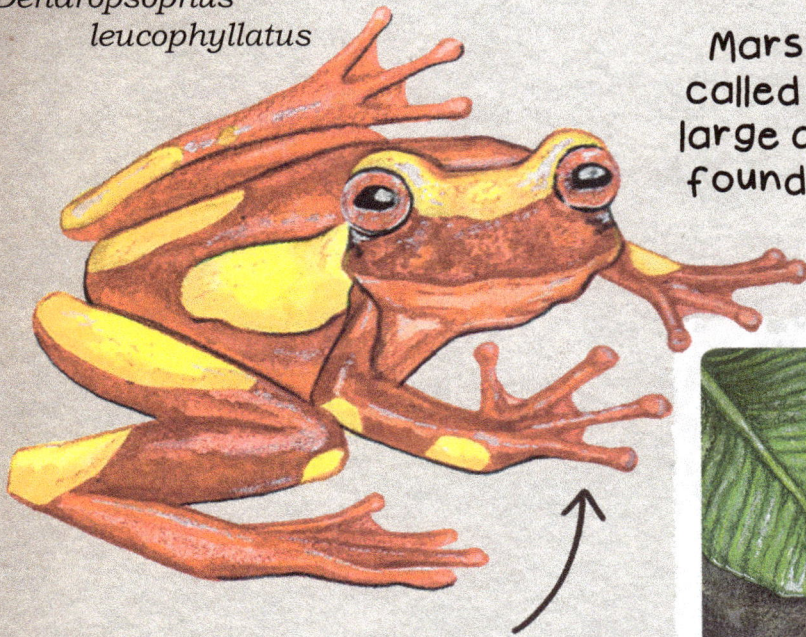

Marshes in the Amazon are often called "Floating Meadows" due to the large amount of free floating plants found on the surface of the water.

Did you know? Not all tree frogs live in the trees but something they all have in common is their toe pad shape, and an extra bone in their toes to help them climb!

Clown tree frogs live in South America throughout Bolivia, Brazil, Colombia, Ecuador, French Guiana, Guyana, Peru, and Suriname.

Least Concern: Their habitat is large and the population is wide spread.

LC   NT   VU   EN CR EW EX

D is for Dwarf Pixie Frog!

Dwarf pixie frogs are the miniature version of the African bull frog species. Their larger counterparts are the second largest species of frog in the entire world. They can get up to 10 inches long and weigh up to four pounds.

Also called the edible bull frog, local indigenous peoples find them to be a tasty treat.

These frogs live throughout grassland areas. If there are any puddles to be found they will sit in them.

During dry seasons, they burrow in the ground to cool down and retain moisture.

Found in savannah regions south east of Nigeria, Somalia, Port Elizabeth, and Angola, Africa.

Least Concern: Their population is thought to be decreasing due to collection for human consumption.

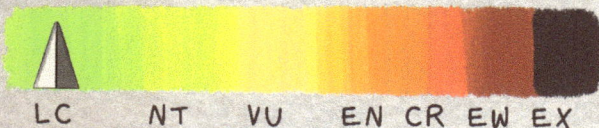

LC   NT   VU   EN CR EW EX

Pyxicephalus edulis

E is for Emperor Newt!

Emperor newts are a rough skinned newt with rough, bumpy skin and ridges down their back.
Their tail is flat and similar to the shape of a paddle.
Other names for this newt include the Mandarin newt, and Mandarin salamander.

These newts prefer slow moving streams and pools of water in subtropical forest areas.

Salamanders and newts are poisonous?!

Most brightly colored, rough skinned salamanders and newts like the emperor newt contain a paralyzing poison called tetrodotoxin or TTX.

This poison is found in the skin, muscles and internal organs. It is the same poison the puffer fish contains.

The emperor newt is only found in the Yunnan Province of China. It lives in the mountains along the Nu, Lancang, and Yuan Rivers.

Near threatened: Facing habitat loss, and removal from habitat for use in medication.

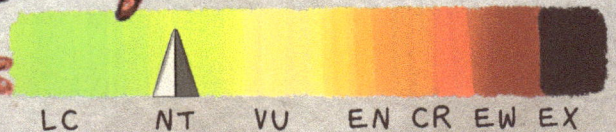

LC   NT   VU   EN CR EW EX

Tylototriton shanjing

F is for Frosted Flatwoods Salamander!

Frosted Flatwoods Salamanders get their name from flatwood forests they live in. These forests are made up of hard clay-like soil and trees with shallow root systems. Due to the hard soil, these forests have poor drainage and often flood during spring, creating temporary ponds.

Amphibians must lay their eggs in water, so this type of forest is perfect for them.

Salamanders: a main group of amphibian characterized by their long tail like a lizard and smooth skin like a frog.

There are many kinds of salamanders. Those that live most of their life on land have claw-like toes that they use to dig burrows.

Some other types of salamanders include olms, newts, and sirens!

Ambystoma cingulatum

Vulnerable: Their habitat is in decline due to deforestation for agriculture.

LC   NT   VU   EN CR EW EX

Lives throughout the Southern Atlantic coast of the United states from South Carolina to the Florida pan handle.

G is for Great Crested Newt!

Great Crested Newts are known for the dramatic jagged crest that males grow during spring breeding season. These newts will "dance" waving their crested tale around to impress females. As unique as a finger print, each newt has its own black spotted pattern on its orange belly. Due to its bumpy skin, this newt has been nicknamed the "warty newt."

All newts are salamanders but not all salamanders are newts! The term "newt" refers to a type of salamander that often spends a portion of its life in water. It will adapt a paddle like fin on its tail and webbed toes for better swimming.

The great crested newt is only found throughout Northern and Central Europe.

After breeding season, great crested newts will absorb the crest on their back as nutrition!

Triturus cristatus

Least Concern: Their population is closely watched under protection by European law.

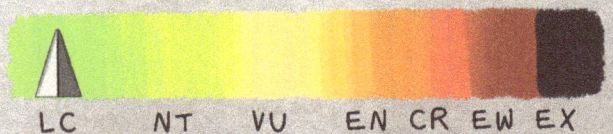

LC   NT   VU   EN CR EW EX

H is for Hellbender Salamander!

Hellbender salamanders are the third largest species of salamander in the world. They can grow to over two feet long and weigh over four pounds. Their closest relatives are the Giant Salamanders of China and Japan. Hellbender's require swift-moving, unpolluted streams and rivers to live in. The quick moving water provides higher oxygen levels. They are one of few salamanders that live underwater their entire lives.

The eastern hellbender is found in the United States in parts of New York, west to Illinois, south to northeastern Mississippi and parts of Alabama and Georgia.

As "ambush" predators they hunt like eels, waiting for their prey to swim by while they stay concealed in between large, flat rocks.

Loyal to their rocks, these salamanders have been found under the exact same rocks year after year.

Fossil records have shown these salamanders lived over 160 million years ago!

Near threatened: This species faces habitat loss and degradation.

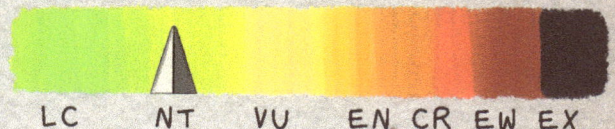

Cryptobranchus alleganiensis

LC   NT   VU   EN CR EW EX

I is for Iberian Painted Frog!

Iberian painted frogs are part of a primitive family of amphibians called the Discoglossidians. They have a unique disc-shaped tongue that they are unable to move since its attached at the base. This means, it can not extend its tongue to catch prey as we commonly see with other frogs.

Iberian painted frogs are known for their skin patterning. They are covered in large dark spots, with light edges, and bright bands down their backs that look as if they were hand painted.

Found in temperate forests, swamplands, grasslands marshes, rivers, shrubby vegetation, sandy shores and pretty much anywhere near water.

Endemic to the Iberian Peninsula occupied by Portugal and Spain.

Least Concern: Wide distribution but likely to be declining enough to qualify for a more threatened listing.

LC   NT   VU   EN CR EW EX

Discoglossus galganoi

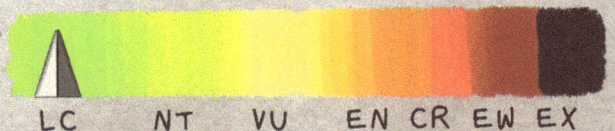

J is for Java Flying Frog!

The java flying frog cannot fly as their name suggests, but are known to jump and glide down through the tree canopies to breeding pools below. A canopy species, this frog is considered rare because its not often seen and hard to locate in tree cover.

Rhacophorus margaritifer

Like many other tree frogs, they are thought to only descend to ground level to lay eggs in streams.

The frogs create foam nests high up above pools and ponds in the rain forest.

This process has been described as churning mucous!

Also known as the Javan tree frog, this frog is native to Java, Indonesia but can also be found throughout Thailand, and Malaysia.

Least Concern: Their population is decreasing due to tree farming and habitat loss, as this frog depends highly on forest canopies.

LC   NT   VU   EN CR EW EX

K is for Koa Toa Island Caecilian!

Koa Toa Island caecilians are apart of the most rare group of legless amphibians. Caecilians are worm-like creatures live in tunnels and burrows they create underground. So much is still unknown about this species because they are rarely seen. While most live under ground, some are found to live in water. Scientists suspect there to be over 200 kinds of these amphibians, and some are known to get up to five feet long.

Caecilians have dozens of needle-sharp teeth lining their mouths. Teeth form in rows similar to the mouth of a shark!

Funnily enough, they don't use their teeth to chew, as they swallow food whole.

Caecilians like tropical forests with loose soil, leaf litter, and rivers or streams nearby.

Caecilians range from Central Africa, Southeast Asia, and Mexico to Argentina.

Ichthyophis kohtaoensis

Least Concern: Their population is stable.

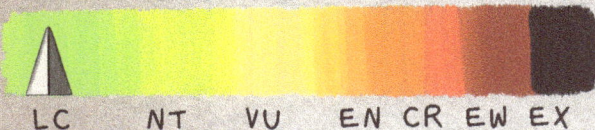

LC   NT   VU   EN CR EW EX

L is for Luristan Newt!

The luristan newt, (also known as the Iranian harlequin newt) is a colorful, smooth-skinned salamander.

These newts live in the dry, arid climate of the Middle East where pools of water and streams can dry up for months.

To avoid the intense heat of the desert conditions, the luristan newt will burrow underground for a period of Aestivation.

With a groovy looking patterned skin, the bright orange and black colors trick predators into thinking they are poisonous.

The luristan newt only lives in three spring-fed streams in the Southern Zagos Mountains of Iran.

Vulnerable: These newts are illegally caught and sent out of their natural habitats for sale in the pet trade. Other reasons include loss of habitat from drought and damming of streams.

LC NT VU EN CR EW EX

Neurergus kaiseri

M is for Mud Salamander!

The mud salamander is a smooth skinned spotted salamander that ranges from a bright, red orange to brown. Often confused with the red salamander, the mud Salamander will have brown eyes rather than gold and slightly shorter snouts!

Just like the luristan newt, this salamander has evolved to have a bright red coloring that tricks predators into thinking they're poisonous.

Hoplophryne uluguruensis

These salamanders get their name from the muddy swamps, streams and slow moving waters they live in.

There are multiple sub-species of the mud salamander. Overall they live throughout the southwestern coastal states of United States including Tennessee, North Carolina, South Carolina, and Georgia.

Least Concern: Their population is stable.

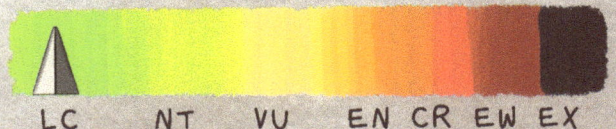

LC   NT   VU   EN   CR   EW   EX

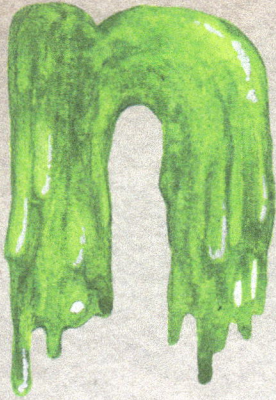

N is for Northern Slimy Salamander!

The northern slimy salamander is a smooth-skinned salamander that gets its name from a glue-like substance that covers its skin when disturbed by predators! The "slime" substance comes from glands in the salamanders skin and is difficult to remove. When predators encounter the substance, it reduces jaw movement making them less of a threat.

Fun Fact: The scientific name of the northern slimy salamander is

Plethodon glutinosus

plethore is Greek, meaning "full of teeth" referring to the large number of teeth this salamander has! glutinosus is Latin meaning "full of glue or very gluey"

They are found in a large range of the United States throughout Texas, Florida, Missouri, Illinois, New York and Connecticut.

Least Concern: Most states have stable populations. Populations in Connecticut have been reported as threatened.

LC   NT   VU   EN CR EW EX

Plethodon glutinosus

O is for Orange Thighed Tree Frog!

The orange thighed tree frog have bright orange colored thighs and underbellies as the name suggests. Their orange eyes match the color of their thighs, this is the easiest way to tell them apart from the Australian red eyed tree frog.

The orange thighed tree frog is similar to the Australian red eyed tree frog which you might recognize as one of the mascots of the famous Rain Forest Cafe!

They live in tropical rainforests of Queensland, Australia.

Least Concern: Their population is stable.

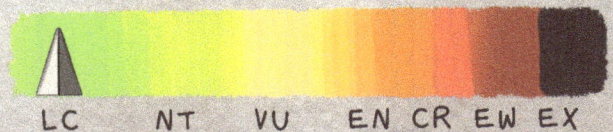

Litoria xanthomera

LC   NT   VU   EN CR EW EX

P is for Pebble Toad!

The pebble toad is one of the most extraordinary amphibians that will no doubt make up for the lack of toads included in this book!

As we have learned, amphibians all have their own ways of avoiding predators like camouflage or coloration. This bumpy, small black toad has one of the most unique strategies... their Rock and Roll! escape method.

When threatened, they curl into a ball, and drop, or roll away!

Due to its lightweight body, it doesn't get hurt when bouncing down a rocky mountainside.

This defense is important because this toad can not hop, or swim!

These toads are endemic to the Guiana Highlands in Bolivar State, Venezuela.

Oreophrynella nigra

Vulnerable: Habitat loss is the main threat for this toad, as they are only found in the Guiana Highland Mountain Range.

LC   NT   VU   EN CR EW EX

Q is for Quacking frog!

The quacking frog is a short, flat, ground-dwelling frog. These frogs have distinctive red coloring along their inner legs, earning them their alternative name: The red-thighed froglet.

The frogs scientific name *Crinia georgiana tschudi* comes from King George Sound, where the species was first discovered. These frogs prefer low-lying moist areas within forests.

Quacking frogs are most well known for their loud call that resembles the "quack" of a duck. Male frogs call and respond to mimic each others' notes.

These frogs have been known to be cannibalistic, and will eat smaller frogs.

They are found only in the southwest region of Australia from Cape Le Grand to Gingin.

Least concern: Their population is stable.

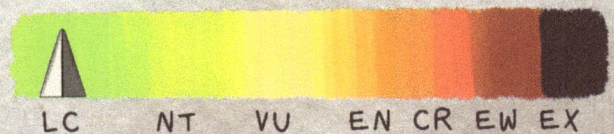

LC   NT   VU   EN CR EW EX

Crinia georgiana

R is for Rococo Toad!

The rococo toad is sometimes referred to as cururu toad, or schneider's toad. This toad is very large at seven to ten inches. Rococo toads inhabit a variety of habitats, most commonly in open, urban areas. They're also known to be found in residential yards and gardens.

Toad Toes!

Front feet: un-webbed with four toes

Back Feet: webbed with five toes

Most toads will have short legs because they crawl rather than jump like frogs.

This toad is able to inflate its body when threatened; making it difficult to swallow by predators! Their toxic skin glands cause trouble in residential spaces where pets that come into contact will be rushed to the vet from paralysis, and possible death.

They're found in Northern Argentina, Paraguay, Uruguay, Eastern Bolivia, and Eastern and Southern Brazil.

Least Concern: Their population is increasing.

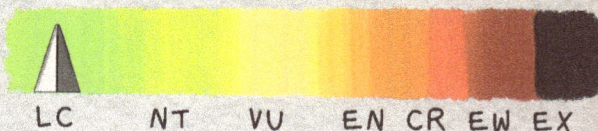

Rhinella schneideri

LC   NT   VU   EN CR EW EX

S is for Solomon Island Leaf Frog!

The Solomon Island leaf frog is a unique, geometrically shaped frog that resembles its habitat! This frog species varies in color from green, brown, yellow, and orange- just like fallen leaves. As nocturnal ambush predators, their camouflage allows them to remain unnoticed before attacking their prey. Other names for this frog include the Solomon Island eyelash frog and Gunther's triangle frog.

They live in the Solomon Islands, Papua New Guinea, Buka Island and Bougainville Island.

These frogs are one of few frog species that don't have a tadpole stage. Their offspring hatch from eggs as fully developed frogs!

These frogs make a loud call that has been said to sound like a dog's bark!

Least Concern: Pet trade is a risk for the frog. The government of the Solomon Islands is considering passing laws to stop the collection of this frog.

LC   NT   VU   EN CR EW EX

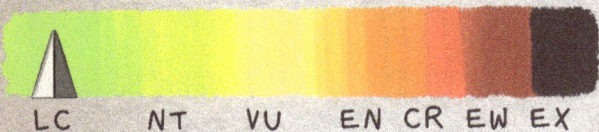

Anaxyrus canorus

T is for Tiger Salamander!

Tiger salamanders are known for their markings that resemble- you guessed it- a tiger's stripes! Their pattern can vary throughout this salamanders geographical range. They are the largest land-dwelling salamander on Earth.

Tiger salamanders dig their own burrows near slow moving waters.

Other salamanders are thought to inhabit the burrows of small mammals or reptiles rather than dig their own.

Tiger salamanders generally live long lives, often up to 16 years in the wild!

The most wide-ranging species of salamander in North America, they are found throughout Mexico, Canada, and the United States.

Least Concern: Their population is stable.

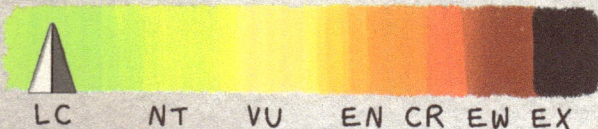

LC   NT   VU   EN   CR   EW   EX

Ambystoma tigrinum

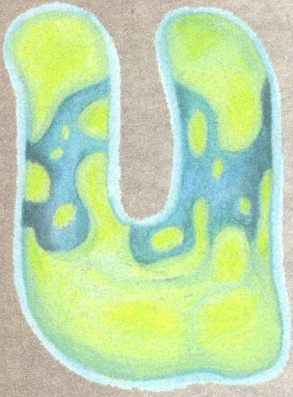

U is for Ulu Guru Blue Bellied Tree Frog!

The unique greenish, blueish coloration of this tree frog is eye catching!

These frogs are covered with white dots and rings to resemble the wet decaying banana leaves they are found on.

Frog Feet!

Front feet: webbed with four toes

Back Feet: webbed with five toes

Not all frogs have webbing on each foot. This trait is found primarily on frogs that spend most of their time in, or near water.

These frogs live only in the Eastern Arc Mountain Range of Tanzania.

Endangered: This species only occurs in the specific Eastern Arc mountain range. At risk by habitat loss.

LC   NT   VU   EN CR EW EX

Hoplophryne uluguruensis

V is for Vietnamese Mossy Frog!

The Vietnamese mossy frog has an interesting camouflage technique, resembling a clump of moss to blend into its habitat of moss-covered, flooded limestone caves. This species is semi-aquatic and spends time in floating plants, and between rocks in the water.

Fun Facts:
There are 11 other bumpy species of this frog throughout South east Asia. Some look like bark-others-like the warty tree frog look like bird poop!

These frogs can throw their vocal calls up to 13 feet, making them hard to find in the wild!

Play dead: these frogs roll into a ball when frightened!

The Vietnamese mossy frog lives in Northern Vietnam and China.

With its tubercules, or small spiny warts all over the frogs body-this frog physically resembles the texture moss.

Least concern: The population is in decline due to deforestation and loss of habitat.

LC   NT   VU   EN CR EW EX

Theloderma corticale

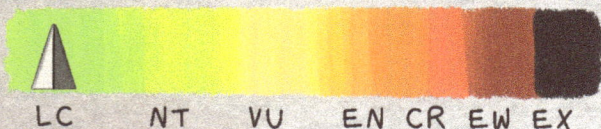

W is for White's Tree Frog!

The White's tree frog is a common frog that prefers moist forest environments, but are adaptable to dryer climates. This frog is known among residents of Australia, as they find moisture wherever they can. They have been seen sticking to public toilets, porches and water tanks! When dry seasons hit, these frogs can produce and cover their bodies with a milky substance called "caerviein" which helps them stay hydrated.

The White's tree frogs eyes have pupils that sit sideways, unlike most other frogs!

When threatened, this species will shout an ear piercing vocal call!

Ranoidea caerulea

These frogs have giant toe pads that help them stick high up on leaves that collect rainwater.

Native to Australia and Southern New Guinea and Indonesia.

Least Concern: Their population is stable.

LC   NT   VU   EN CR EW EX

X is for Xenopus!

The African clawed frog (whose scientific name is *Xenopus laevis*) is a very unique looking frog named for its long, "claw-like" fingers. These claws help the frog tear apart their underwater prey, and allow them to push food into their mouths. The frogs' mottled brown and green coloring helps them camouflage from predators in shallow waters.

Fun Facts! In the early 1900's, African clawed frogs were used worldwide for pregnancy testing. The frogs produce eggs in response to a chemical in the urine of a pregnant person.

The African clawed frog can burrow in the mud and hibernate for up to a year!

Xenopus laevis

Webbed back feet are great for swimming, but make this frog clumsy on land; crawling rather than hopping.

Its found throughout Africa south of the Sahara Desert. This species is fully aquatic and prefers to live in quiet stagnant waters such as ponds, or pools of standing water in Savannas.

LC NT VU EN CR EW EX

Least Concern: The African clawed frog has become a widespread, invasive species since the frogs used for testing were released into the wild after pregnancy tests became readily available.

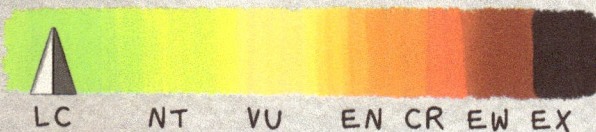

Y

Y is for Yosemite Toad!

The yosemite toad was first discovered in, and named after Yosemite National Park. These toads are only active for a couple months of the year. During the winter they will hibernate in the burrows of small mammals.

A national forest biologist recently discovered that these toads were often risking their lives crossing busy highways during breeding season.

Hoping for a solution to save these toads, a series of small, underground passageways were put in place to help the toads safely find their way to their hibernation habitat.

Holy Cow?

Looking for moisture and shade during the hot months in the meadow, these toads are often found hiding in cattle hoof prints!

The Yosemite toad lives in California from the high elevations of the Sierra Nevada Mountains to Fresno.

Endangered: This species has been in dramatic decline for decades. Some factors include destruction of habitat, chytrid fungus and climate change.

Anaxyrus canorus

LC   NT   VU   EN CR EW EX

Z

Z is for Zhangs Horned Frog!

The Zhangs horned frog, also known as the Zhangs spadefoot frog is a species of leaf-litter frog. Their leaf-like projections over their eyes help them blend into the forest floor!

Found near Zhangmu, China. A port of entry located in Nyalam County on the Nepal-China border, Tibet.

Lives in humid subtropical forest areas near rivers.

Not much is known about this frog, as it is thought to be a new species of *Xenophrys Kuhl*.

Xenophrys zhangi

Zhangmu, China is just north of the Friendship Bridge border crossing. This city has since been evacuated and left a ghost town, due to damage from earthquakes.

Near Threatened: This frogs habitat is in decline.

LC   NT   VU   EN CR EW EX

Amphibians need your help
various ways you and your family can help protect and conserve species

Avoid using pesticides and chemicals: Chemicals such as pesticides, herbicides and fungicides harm wildlife, including insects that amphibians eat. Pouring chemicals down drains or using them outside and polluting water runoff can be detrimental to amphibian habitats.

Leave ground cover and native aquatic vegetation: Keep the fallen wood and leaves on the ground, and plants in the water! These natural elements provide safe shelter, food, and a breeding ground to amphibians.

Reduce your lawn and add native plants, or water: Native plants attract and support insect populations, which amphibians rely on as a food source. Adding a source of water such as a low lying bird bath or pond can provide a habitat for amphibians.

Don't capture animals from the wild: Don't capture wild animals or release pets to the wild. Both have negative impacts such as spreading disease, increasing invasive species, and decreasing wild populations.

If you choose to get an exotic pet, look for one that is captive bred: Captive bred amphibians are bred from a small amount of wild-caught amphibians. Breeding in captivity decreases the demand to have wild-caught amphibians as pets. Any wild-caught amphibians used to breed are never sold. Captive breeding can help stabilize and increase a species population.

Do your research and be prepared: Owning an amphibian (just like any pet) is a big responsibility. Amphibians in captivity live an average 2-20 years and some are known to live up to 70 plus years.

Donate: Research trustworthy organizations supporting rainforest, wet-land and amphibian conservation, such as rainforesttrust.org and amphibians.org

For more information on caring for amphibians in the wild and as pets, you can visit epa.gov and joshsfrogs.com

Clown Tree Frog
Dendropsophus leucophyllatus
| 1-1.5" |

Java Flying Frog
Rhacophorus margaritifer
| 1.5-1.8" |

Emperor Newt
Tylototriton shanjing | 4.5-6" |

Northern Slimy Salamander
Plethodon glutinosus
| 4.75-7" |

Frosted Flatwoods Salamander
Ambystoma cingulatum
| 5" |

Iberian Painted Frog
Discoglossus galganoi
| 2.9" |

Axolotl
Ambystoma mexicanum | up to 14" |

Made in U.S.A.

1 2 3 4 5 6 7 8

Solomon Island Leaf Frog
Cornufer guentheri
|2.5-3.5"|

Pebble Toad
Oreophrynella nigra
|.62-.90"|

White's Tree Frog
Ranoidea caerulea
|3-4.5"|

Yosemite Toad
Anaxyrus canorus |2-3"|

Mud Salamander
Pseudotriton montanus
|6-8"|

**Ulu Guru
Blue-bellied Tree Frog**
Hoplophryne uluguruensis
|1.18-1.5"|

Blue Poison Dart Frog
Dendrobates tinctorius
"azureus" |1.5-2"|

Tiger Salamander
Ambystoma tigrinum |6-8"|

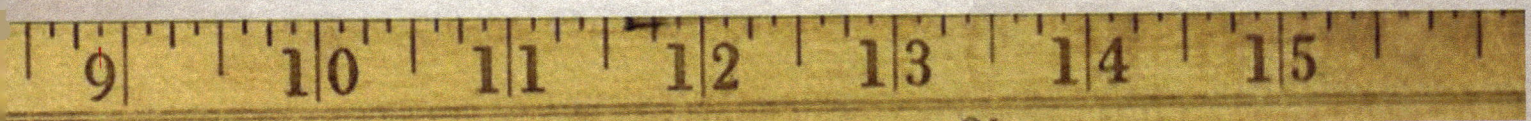

9 10 11 12 13 14 15

Zhang's Horned Frog
Xenophrys zhangi
| 1.7" |

Vietnamese Mossy Frog
Theloderma corticale
| 3.5" |

African Clawed Frog
Xenopus laevis | 2-5" |

Dwarf Pixie Frog
Pyxicephalus edulis
| 5.5-10" |

Orange Thighed Tree Frog
Litoria xanthomera
| 2.2" |

Rococo Toad
Rhinella schneideri
| 7-10" |

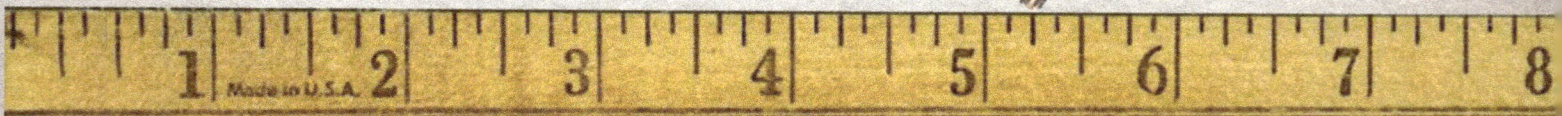

1   Made in U.S.A.   2   3   4   5   6   7

Great Crested Newt
Triturus cristatus
|6.69-7"|

Hellbender Salamander
Cryptobranchus alleganiensis
|12-30"|

Koa Toa Island Caecilian
Ichthyophis kohtaoensis
|7-21"|

Quacking Frog
Crinia georgiana |1.5"|

Luristan Newt
Neurergus kaiseri
|5.11"|

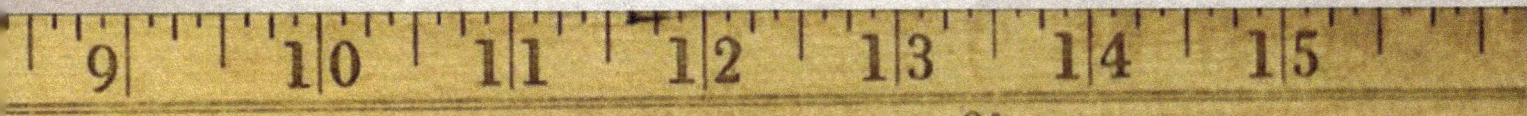

9   10   11   12   13   14   15

Amphibians of the World!

- Axolotl
- Blue Poison Dart Frog
- Clown Frog
- Dwarf Pixie Frog
- Emperor Newt
- Frosted Flatwoods Salamander
- Great Crested Newt
- Hellbender Salamander
- Iberian Painted Frog
- Java Flying Frog
- Koa Toa Island Caecilian
- Luristan Newt

Mud Salamander

Northern Slimy Salamander

Orange-Thighed Tree Frog

Pebble Toad

Quacking Frog

Rococo Toad

Solomon Island Leaf Frog

Tiger Salamander

Ulu Guru Blue-bellied Frog

Vietnamese Mossy Frog

White's Tree Frog

Xenopus: African Clawed Frog

Yosemite Toad

Zhang's horned Frog

GLOSSARY
did you know?

Aestivation: a period of inactive time where an animal will hibernate to avoid extreme heat

Canopy: the upper most portion of trees made up of branches, leaves and stems

Camouflage: to hide, blend into, or disguise

Captive: to be caged, restrained, no longer in natural habitat

Conservation: to prevent loss, or wasteful use of something

Chytrid Fungus: a disease spread between amphibians that attacks their skin; can be deadly

Dorsal fin: a fin located on top or on the back of animals that helps direct through water

Endemic: a plant or animal that is native, or restricted to a certain area

Ethical: what is right or wrong- relating to rules or standards

Hibernation: a period of inactive time where animals escape from the extreme cold

Larval: a young or immature stage of growth

Neoteny: where an animal keeps its larval features into adulthood

Metamorphosis: the process of an amphibian transforming from an immature stage to adult

Rain forest: an environment with heavy tree cover and moisture reliant plants

Sustainable: taking care with use of our resources in order to protect nature long term

Wild Caught: an animal removed from their natural habitat

Color your own Frog and Salamander Print.
How might they Camoflauge into their surroundings? Cut out and frame!

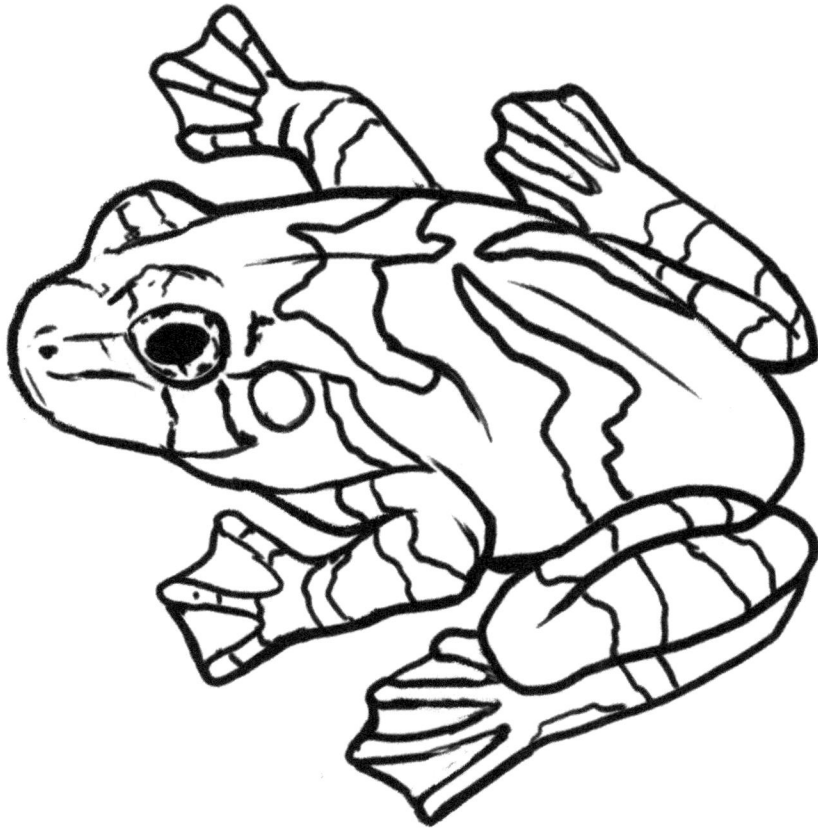

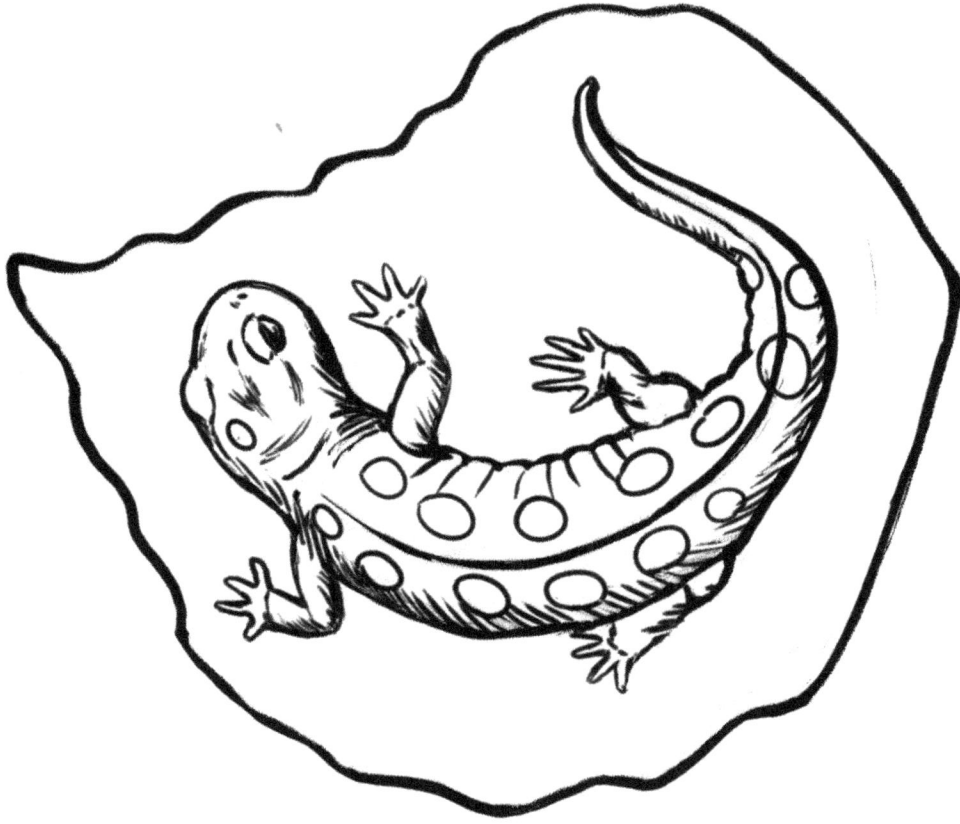

Free space for drawing your own amphibians and placing stickers!

www.ingramcontent.com/pod-product-compliance
Lightning Source LLC
Chambersburg PA
CBHW040303100426
42811CB00011B/1344